Sprout

The Plant Files

Discovery Channel School Science Collections

© 2000 by Discovery Communications, Inc. All rights reserved under International and Pan-American Copyright Conventions. No part of this book may be reproduced in any form or by any electronic or mechanical means, including information storage devices or systems, without prior written permission from the publisher. For information regarding permission, write to Discovery Channel School, 7700 Wisconsin Avenue, Bethesda, MD 20814. Printed in the USA ISBN: 1-58738-001-3

1 2 3 4 5 6 7 8 9 10 PO 06 05 04 03 02 00

Discovery Communications, Inc., produces high-quality television programming, interactive media, books, films, and consumer products. Discovery Networks, a division of Discovery Communications, Inc., operates and manages Discovery Channel, TLC, Animal Planet, Discovery Health Channel, and Travel Channel.

Writers: Jackie Ball, Denise Vega, Uechi Ng, Kimberly King, John-Ryan Hevron, Judy Gitenstein, Judi Christy, Kathy Feeley, Susan Wernert Lewis, Scott Ingram, Monique Peterson, Katie King. **Editor:** Katie King. **Photographs:** Cover, sprouting plants, ©Dwight Kuhn; p. 3 and p. 17, George Washington Carver, ©Brown Brothers, Ltd.; p. 5, strangler fig, ©Gregory G. Dimijian/Photo Researchers, Inc.; p. 10, apple picker, ©Stephanie Maze/CORBIS; p. 11, Johnny Appleseed, ©CORBIS-Bettmann; p. 12, Henry David Thoreau, ©Brown Brothers, Ltd.; p. 15, bladderwort, ©Dwight R. Kuhn; sundew, ©Bill Beatty; monkey cup and Venus flytraps, ©PhotoDisc; p. 16, George Washington Carver, ©Bettmann/CORBIS; p. 17, George Washington Carver in field, ©Brown Brothers, Ltd.; p. 18, bat pollinating cactus, ©Merlin D. Tuttle/Bat Conservation Intl.; bee on blossom, ©PhotoDisc; p. 19, hummingbird, ©Anthony Mercieca/Photo Researchers, Inc.; bee on orchid, ©Nuridsany & Perrenou/Photo Researchers, Inc.; p. 20, tree fern, ©Terry Whittaker/Photo Researchers, Inc.; fiddlehead fern, ©Farrell Grehan/Photo Researchers, Inc.; p. 21, wisk fern, Kathy Merrifield 1997/Photo Researchers, Inc.; spikemoss "resurrection plant," ©David Sieren/Visuals Unlimited; filmy fern, ©Kjell B. Sandved/Visuals Unlimited; pp. 24–25, all photos courtesy Suzanne Nelson; p. 27, bat, ©Merlin D. Tuttle/Photo Researchers, Inc.; cactus flower, ©Doug Sokell/Visuals Unlimited; p. 30, *Amorphophallus titanum*, ©Prance/VU; p. 30, girl, ©Comstock; p. 31, *Little Shop of Horrors*, ©Everett Collection; all other photographs, ©Corel. **Illustration:** p. 29, wilderness scene, Leslie Evans. **Acknowledgments:** p. 14, THE DAY OF THE TRIFFIDS, by John Wyndham. ©Doubleday and Co., Inc., 1951. Reprinted with permission.

Sprout

CONTENTS

The Great Producers

Plants are all around us. But they're not the stagnant creatures you might think! Remarkably, plants spend their everyday lives doing the same things most animals do. Plants fight for survival—for turf, light, and sometimes mates. But unlike animals, plants have the unique ability to make their own food. The middlemen between us and the Sun, plants take the Sun's energy and convert it into food inside their cells.

Plants are pretty busy. In ecosystems around the world many diverse species have ingenious ways of coping with their environments. How plants live in their environments has much to do with their survival strategies and adaptations.

Come along with SPROUT and Discovery Channel and get to know your friends the plants. You won't believe some of the amazing things plants do.

The Plant Files

Plants. 4
At-A-Glance Why care about plants? Because without them, we couldn't exist!

Vegging Out. 6
Q & A Disproving the idea that it might be a vegetable, the tomato clears up some rumors about itself and its friends.

Plant Planet 8
Almanac Not all plants are alike! The plant species in our world are diverse. Find out about the major groups and what makes them all so different.

An Apple…A Year 10
Timeline The year in the life of an apple tree is complicated. Learn about the growth cycle of an apple tree and how farmers cultivate one of America's favorite fruits.

"Faith in a Seed" 12
Eyewitness Account Uncover what Henry David Thoreau meant when he talked about seed dispersal and the changing lands around Concord, Massachusetts.

Mean and Green 14
Scrapbook Who says plants are dangerous? After all, plants need food, too. Find out how some plants get fed.

The Plant Doctor 16
Heroes George Washington Carver was the scientist who came up with more than 300 uses for the peanut. He was also known around the world as a gentle teacher who loved plants.

To Bee or Not to Be 18
Picture This The way flowers look has a lot to do with how they pollinate. There are many players in the process of pollination. Learn who they are and what they do.

Older Than the Dinosaurs 20
Amazing But True Plants without seeds were the first living plants on Earth. Find out how their modern relatives get around.

Meet the plant doctor; see page 17.

Around the World in 80 Bites 22
Map Many of our major food crops originated in some far-off places. Take a trip around the world to see where your favorite foods may have come from.

Saving Seeds 24
Scientist's Notebook Native Seeds/SEARCH runs a seed bank, serving the Native American community by preserving seeds and traditional farming methods in the Southwest.

Livin' Large 26
Virtual Voyage What does it take to live in the desert? Become an organ-pipe cactus and see for yourself.

The Suspicious Specimens 28
Solve-It-Yourself Mystery Camp counselor Herb Greene's scouts are on a mission to track down plants. But some of the plants they've gathered just don't add up. Can you figure out which scout is a real thorn in Greene's side?

Wild Plants 30
Fun & Fantastic From the biggest to the stinkiest, plants come in all personalities, and many have their own stories to tell.

Final Project
Classroom Garden 32
Your World, Your Turn The United States is more diverse in plant life than we ever realized. Even so, many plants are threatened. Discover the plants that grow around your school. Find out what makes them all so special and why it's important to keep plants around.

Plants

AT-A-GLANCE

You're on the tundra, on a vast, windswept plain north of the Arctic Circle. There are no trees, no fields of flowers—it seems as if there are no plants at all. But the plant kingdom here is actually diverse and determined. Plants come in an amazing variety of forms, and grow and even flourish in some of the most extreme environments on Earth. If you look closely at this tundra, you'll see mosses, lichen, cotton grass, willows, and wildflowers that cling low to the ground. Just as these plants have adapted to survive in their environment, so have plants in other places around the world.

Cotton grass in the arctic

tundra

Deep inside the rain forest, plants have adapted to an environment with lots of moisture, but poor soil. Seeking light, ivy scrambles up tree trunks to reach the top of the rain forest canopy. The strangler fig survives at the expense of the tree on which it lives. It squeezes the trunk of its host, choking it to death. Elsewhere, in a woody forest, mistletoe steals its host's water and nutrients. The "living stone" plant of southern Africa survives harsh desert conditions by camouflaging itself as a stone. That way animals won't eat it.

Yet despite diversity, plants are the source of energy for nearly all life on Earth, and all plants have one thing in common: the remarkable ability to convert sunlight into food. Chlorophyll, which makes plants green, harnesses energy from sunlight. The plant uses that energy, along with the water it has collected through its roots and the carbon dioxide through its leaves, to create a sugar to use as food. This process is called photosynthesis, and only plants can do it. Imagine if your body could make its own lunch just by standing in the sunshine! And the by-product of photosynthesis—the waste product that a plant gives off—is the oxygen that we breathe.

We all need plants. The food you eat, the cotton shirt on your back, the paper on which you write, the fuel that drives your car, the medicines you take, and even the book you're reading now—all these things and many more come from plants.

Looking up into a strangler fig after the host tree is dead and rotted away.

rain forest

Vegging Out

Tomato tells all

Q: You're a tomato, star of salads, sandwiches, soups, and sauces. Bursting with vitamins. Fat-free. No cholesterol, low sodium...

A: All true.

Q: So, you really stand out from other plants.

A: Well, I am unique, but I actually do what all plants do. We use the Sun's energy, along with water and carbon dioxide to create food. It's called photosynthesis. Basically we provide all the energy for nearly all life on Earth.

Q: That's cool. So, what's it like to be a vegetable?

A: Oh, I'm not a vegetable.

Q: But that's ridiculous. Of course you are. You're the same kind of food as all our other favorite veggies—eggplant, squash, green peppers.

A: I don't know how to break this to you, but they're not vegetables, either.

Q: But then what *are* all of you?

A: We're fruits.

Q: Fruits? As in peaches and plums? Mangoes and melons? Lemons and limes?

6 DISCOVERY CHANNEL SCHOOL

A: Bananas and blueberries. Yes, indeed. In fact, in some ways I am identical to a blueberry, only bigger. And, um, not blue.

Q: This is too much! You mean all my life I've been putting slices of fruit on my sandwiches . . . eating spaghetti with meatballs and fruit sauce. Yuck!

A: Oh, get over it. Fruit, vegetable—what difference does it make? They're just words. Names.

Q: Well, you started it, so you tell me: What difference *does* it make?

A: It has a lot to do with what part of the plant you eat. If you eat the stem, leaf, or root, we call it a vegetable. Think cabbage or lettuce leaves. Roots, as in carrots or radishes. Leaf stalks of celery, or underground stems, called tubers, like potatoes. Usually the part you eat is the place where nutrients are stored. Asparagus stores sugars and starches in its flowering stems, for instance.

Q: So how are fruits different?

A: Fruits often store sugars and starch, too, but they contain seeds. They're part of a plant's reproductive system. Fruits also can be firm and fleshy (like yours truly), or hard, like a pea pod. They can be juicy or dry. Some fruits you eat, some you don't. I wouldn't advise eating a walnut shell, for instance. Ouch.

Q: You mean walnuts are fruits, too?

A: That's right. And pecans and almonds and acorns and hazelnuts—when the shells are on, of course. They all contain seeds, and that's one thing that makes a fruit a fruit. Grains are fruits, too. Oats, wheat. They're the dry fruits of cereal grasses. The fruit wall comes off during winnowing.

Q: This is unbelievable! Fruits are everywhere.

A: For a very good reason.

Q: Which is?

A: Which is that all fruits develop from flowers, and there are a lot of flowering plants in this world. In fact, flowering plants make up the biggest plant group on Earth—angiosperms. There are about 250,000 different kinds.

Q: I'm impressed.

A: You should be. We're a very diverse and talented group. Take our flowers, for instance. Each blossom may have both male and female reproductive parts. A male part, called the stamen, produces pollen—you know, that yellow dust that makes some people sneeze in the springtime? A female part, called the pistil, has to be dusted with that pollen before a fruit can form.

Q: How does that happen?

A: Nature lends a hand—or a foot. Sometimes small animals spread the stuff around with their feet. Sometimes birds and insects get dusted when they stop to sip nectar from a flower, and sometimes it floats along on the wind. But once it reaches the pistil, it's likely to stay put.

Q: Why is that?

A: Pistils are specially built to trap that precious dust. They're either sticky or have lots of little hairs that can hold onto it. Once the pollen is on the pistil, it travels down inside the flower to the plant's ovary, which is a kind of pod. And once there, it fertilizes an egg to make embryos. Baby plants. A covering grows around them until you've got. . . .

Q: Seeds?

A: Exactly. Up above, the flower parts fall off, but down below the ovary keeps swelling and growing. Before long, it's mature. The fruit is ripe. And because angiosperms are so diverse, we produce a huge variety of fruits. Fleshy fruits like blueberries, bananas, peaches, kiwis, and cucumbers. Dry fruits like peas, nuts, and grains. Fruits like raspberries are clusters that grow from separate ovaries of one flower. Fruits like pineapples grow from the ovaries of many flowers in a cluster.

Q: You know so much about it that I guess it must be true. You really are a fruit.

A: Don't look so disappointed. In botanical terms, I am a berry, and berry proud of it. Grocery stores can display me next to the beets, but I don't care. Call me a fruit, call me a vegetable, but whatever you do, eat plenty of plants. No matter what you call us, we're good for you.

Q: OK, OK. You've convinced me. Just don't expect me to start ordering. . . .

A: Blueberry pizza.

Activity

PLANT PARTS Go to a local nursery, gardening center, or your own yard or park and observe a flowering plant. Draw the flower in detail and analyze its structure. Try to identify the parts of the plant. Use a field guide to help you.

ALMANAC Plant Planet

So far, scientists have identified some 265,000 species of land plants, from mosses to orchids. About 90,000 of these live in rain forests, and scientists estimate there are at least 30,000 more species they don't know about. Most of these probably live in rain forests, too. Read on for some more amazing facts and figures.

Putting Plants in Their Place
Here's how scientists organize land plants.

Soft and Springy
Group name: Mosses and moss allies
How many species: 10,000
Key ID points: small; cushiony; no flowers; no roots or stems
Examples: moss, hornworts, liverworts
Favorite hangouts: damp spots and bogs; grows on rocks and other plants

Moss growing on dead trees

Gone with the Wind
Group name: Gymnosperms
How many species: 800
Key ID points: cones; needle-like or scale-like leaves; no flowers
Examples: evergreen trees (pine, fir, spruce, cedar); some deciduous trees (larch, ginkgo, bald cypress)
Favorite hangouts: forests everywhere

Redwood trees in California

Such Frond Friends
Group name: Ferns and fern allies
How many species: 13,043
Number of fern allies species: 1,043
Key ID points: fern leaves at first resemble the head of a fiddle; no flowers; most have roots and stems
Examples: swordfern, rattlesnake fern, clubmoss, horsetail, quillwort
Favorite hangouts: tropical forests, edges of freshwater sources

Flowering Power
Group name: Flowering plants
How many species: more than 200,000 species
Key ID points: seeds enclosed in a fruit
Examples: bean plants, maple trees, cactuses, roses, buttercups, geraniums
Favorite hangouts: all environments

Prickly pear cactus flower

PLANTS IN PERIL

Plants all over the world are threatened.

Country	Total Species	Species Threatened	Percent of Total
United States	16,108	4,669	29.0%
Jamaica	3,308	744	22.5%
Turkey	8,650	1,876	21.7%
Spain	5,050	985	19.5%
Australia	15,638	2,245	14.4%
Cuba	6,522	888	13.6%
Panama	9,915	1,302	13.1%
Japan	5,565	707	12.7%
South Africa	23,420	2,215	9.5%
India	16,000	1,236	7.7%
Mexico	26,071	1,593	6.1%
Peru	18,245	906	5.0%
Ecuador	19,362	824	4.3%
Brazil	56,215	1,358	2.4%
Colombia	51,220	712	1.4%

Source: International Union for the Conservation of Nature and Natural Resources, 1997

Plant Hall of Fame

Tallest living tree: A giant redwood in California stands about 367 feet (112 m). That's 62 feet (19 m) taller than the Statue of Liberty!

Largest flower: *Rafflesia* can grow up to 3 feet (.9 m) in diameter. It grows in Malaysia. Picture an open umbrella.

3 feet

Smallest plant: *Wolffia* floats on water. A full grown *Wolffia* is usually smaller than this "o." It grows in North and South America.

SHOOTS AND ROOTS

Plants have three major vegetative organs: leaves, a stem, and roots.

Leaves
- Located above ground
- Absorb energy from sunlight
- Photosynthesis (Produce glucose from water and carbon dioxide, using energy from sunlight)
- Store glucose

Stem
- Located above ground
- Support leaves and flowers
- Buds produce new stems
- Photosynthesis
- Store glucose

Roots
- Located below ground
- Absorb water from ground
- Absorb nutrients from soil
- Tubers (potatoes, carrots) store food (starch)

Activity

FIELD TRIP Go to your local park or forest with a camera or sketchbook, pen and pencil, magnifying glass, and envelope for collecting seeds. Examine the plants with your magnifying glass. How many different kinds of plants can you find on your expedition? Draw or photograph the different species. Start a notebook when you get home. Use the information you've gathered to group the plants according to the information on page 8.

THE PLANT FILES **9**

AN APPLE A...

The life of an apple tree is complicated. It goes through many changes in the course of a year.

WINTER
Late December–March

SLOWIN' DOWN 🍎 Winter may be a slow season for an apple tree, but farmers have plenty of work to do. They prune all the trees in the orchard and cut away dead branches. Trimming the tree helps the younger parts of the tree become stronger. With fewer branches in the way, all parts of the tree get the same amount of sun.

EARLY SPRING
April–May

FIRST BUDS 🍎 Once the first leaves and hundreds of small buds start showing up, the tree needs special nutrients. The farmer adds them to the tree's water and soil and plants new trees for future crops. And we mean future—it takes a tree three to four years to grow its first apples, and another six to seven years before it can produce enough apples to harvest.

LATE SPRING
Mid May–Early June

BUSY BEES 🍎 Before a tree can produce apples, its flowers must be pollinated. Farmers release hundreds of thousands of bees into the orchard. Bees eat the pollen and nectar of a flower to survive. The tree benefits, too. When the bee goes to another flower on a different tree, it spreads pollen to a new flower.

Once pollinated, the flowers lose their petals and the base of the flower begins to grow into an apple.

YEAR

Johnny Appleseed
was a real person and folk hero. In the 1800s, Johnny (whose real name was John Chapman) traveled across the United States. He gave apple seeds and saplings to Indians and settlers he met along the way. Wearing no shoes and a tin pot on his head, Johnny was known for his love of nature and his acts of kindness. He's a big reason why there are so many apple orchards in the country today.

SUMMER
June–July

JUNE DROP More apples fall off the trees in June than in any other month. These apples are still small and haven't fully developed. Farmers remove as many as half of the apples that are left on the tree. This may seem like a backward way of doing things, but the small apples aren't ready for harvest. Once the apples are removed, it helps the tree produce a healthy crop for the next season.

LATE SUMMER
August

LEAF ME ALONE! An apple gets its rich color from the Sun. As the remaining apples grow and ripen, farmers cut back the trees, removing leaves so that enough sunlight hits the apples. But it takes 30 to 50 leaves to nourish one apple to grow to its full size, so farmers don't remove too many. The leaves and fruit of the tree need to be protected from moths, aphids, and bacteria, so farmers spray the trees with pesticides.

AUTUMN
September–early November

HARVEST TIME Workers pick the fruit by hand to avoid damaging the apples or the trees. Some apples don't need special treatment, though. Fruit used for cider and applesauce can be picked by machine. After the harvest, apple trees lose their leaves. Winter might be just around the corner, and then it's only a few short weeks until preparations for next year begin all over again.

Activity

FRUIT PARTS Dissect a fruit, such as an apple, pear, or orange. As you take it apart, write a description of each section (skin, seeds, stem) in a notebook. Give details about its structure. Lay the pieces beside each other and sketch them in the notebook.

THE PLANT FILES

Faith in a Seed

Concord, Massachusetts, mid-1800s

You might know him as a writer and philosopher, but Henry David Thoreau was a scientist, too. Thoreau's lab was the fields and forests of his home, where he observed the world around him. He wrote detailed journals and drew sketches of what he saw.

Nothing stays the same in nature. Thoreau saw that over time ponds might become fields, and fields could slowly evolve into forests. Land plants and animals might replace aquatic plants and animals, in the process of succession. Thoreau set out to find out why this happened. He discovered how seeds are dispersed, or spread, and how forests develop.

Here are some excerpts from his unfinished manuscript, *The Dispersion of Seeds*.

It's All About Seeds

During Thoreau's time, some people believed that plants sprang up spontaneously, rather than growing from seeds, roots, or cuttings. Thoreau proved otherwise.

When, hereabouts, a forest springs up naturally where none of its kind grew before, I do not hesitate to say that it came from seeds. Of the various ways by which trees are known to be propagated—by transplanting, cuttings, and the like—this is the only supposable one under these circumstances. No such forest has ever been known to spring from anything else. If anyone asserts that it sprang from something else, or from nothing, the burden of proof lies with him.

Henry David Thoreau (1817–62)

Thoreau understood what modern ecologists now know. Seeds are spread in a great variety of ways, including by wind, water, and animals.

In the fall it [the seed] would be detained by the grass, weeds, and bushes, but the snow having first come to cover up all and make a level surface, the restless pine seeds go dashing over it like an Esquimaux [Eskimo] sledge with an invisible team until, losing their wings or meeting with some insuperable obstacle, they lie down once for all, perchance to rise up in pines. Nature has her annual sledding to do, as well as we. In a region of snow and ice like ours, this tree can be gradually spread thus from one side of the continent to the other.

Little Messengers

Animals also play an important role in seed dispersal. "Hitchhikers" or burrs—seeds with sticky coats, hooks, or spines—will stick to an animal's fur, or to your socks or jeans if you're out for a walk on an overgrown path. Eventually the seeds fall off, and some may later germinate. Birds and other animals eat berries, which contain seeds. The seeds are deposited in the animal's waste, which falls on the ground. Some animals do even more. One day Thoreau spotted a red squirrel burying something in the ground. Once the animal left,

Red squirrel

This meadow could be a forest in 100 years.

he approached the hole and started digging.

I found two green pignuts joined together, with the thick husks on, buried about an inch and a half under the reddish soil of decayed hemlock leaves—just the right depth to plant it. In short, this squirrel was then engaged in accomplishing two objects, to wit, laying up a store of winter food for itself and planting a hickory wood for all creation. If the squirrel was killed, or neglected its deposit, a hickory would spring up. The nearest hickory tree was twenty rods distant.

Change of Scene

Seeds are also spread by floating on rivers, streams, and oceans. Thoreau describes the changes he sees in a pond.

If you dig a pond anywhere in our fields you will soon have not only waterfowl, reptiles, and fishes in it, but also the usual water plants, as lilies and so on. You will no sooner have got your pond dug than Nature will begin to stock it. Though you may not see how or when the seed gets there, Nature sees to it. She directs all the energies of her Patent Office upon it, and the seeds begin to thrive.

As seeds spread, the landscape changes. Nature moves in quickly. Imagine a pond that is becoming dry. Mosses and grasses usually arrive first, and soon you have a field. Other flowering plants and some conifers follow. These attract insects, birds, and other animals. After about five years, the field is filled with shrubs and trees. If left undisturbed for 100 years or so, the field becomes a woodland.

Nature works no faster than need be. If she has to produce a bed of cress or radishes, she seems to us swift; but if it is a pine or oak wood, she may seem to us slow or wholly idle, so leisurely and secure is she. However, Nature is not always slow in raising pine woods even to our senses. You have all seen how rapidly, sometimes almost unaccountably, the young white pines spring up in a pasture of clearing. Small forests thus planted soon alter the face of the landscape. Last year perhaps you observed a few little trees there, but next year you find a forest.

Activity

SEED JOB Seed dispersal helps plants live in diverse ecosystems. Find out about the plants in your area. Analyze the environment in which they live to determine what sorts of adaptations the local plants have for seed dispersal. For example, if you live in an area with many rivers and streams, how do the plants take advantage of this means for seed dispersal?

Mean and Green

When you think of aggressive creatures that go to extremes to get food, protect themselves, and reproduce, plants may not come to mind. But look closer. In unfriendly environments, plants have adapted some "unfriendly" survival strategies.

Don't Bug Me

Some environments have soil that is very low in nitrogen, an essential nutrient for plants. Solution? Carnivorous plants have adapted by capturing insects, which are generally rich in nitrogen. These plants have evolved special organs, like sticky hairs, to attract and capture their insect prey. Once caught, the prey is digested by special enzymes that dissolve the insect. One of the most interesting yet misunderstood plants is the Venus flytrap. Some people buy Venus flytraps thinking that they will be able to grow man-eating monsters! Venus flytraps can only eat bugs that are a third to a half the size of their traps. Bugs that are larger will either escape or won't be digested quickly enough. They'll end up rotting and damaging the plant. Needless to say, humans are in no danger of being eaten by carnivorous plants!

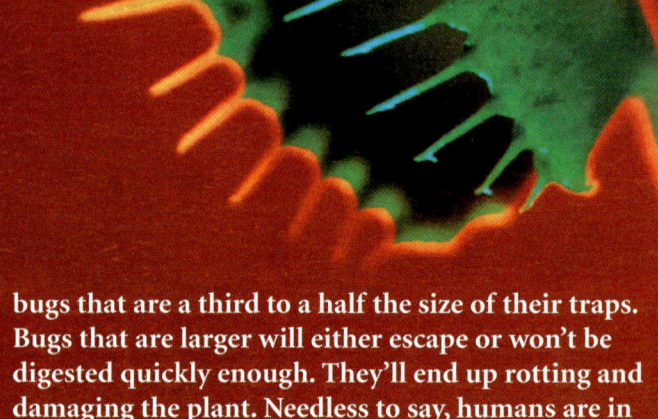

Venus flytrap

Attack of the Killer Triffids

What if plants could walk? And what if triffids, fictional plants, went after people with deadly poison and ate them? Help! This is the setting for *The Day of the Triffids*, a science fiction novel by John Wyndham. Read on.

People were surprised and a little disgusted to learn that the species was carnivorous, and that the flies and other insects caught in the cups were actually digested by the sticky substance there. Also alarming was the discovery that the whorl topping a triffid's stem could lash out a slender stinging weapon ten feet long, capable of discharging enough poison to kill a man if it struck squarely on his unprotected skin.

It was . . . quite a while before anyone drew attention to the uncanny accuracy with which they aimed their stings, and that they almost invariably struck for the head. Nor did anyone at first take notice of their habit of lurking near their fallen victims.

I crossed the bedroom on tiptoes and pulled the window shut sharply. Even as it closed, a sting whipped up from below and smacked against the glass. We looked down on a thicket of triffids standing ten or twelve deep against the wall of the home.

Hair-Raising Secrets of Carnivorous Plants

Gotcha!

Close-up of sundew with lunch

Sundew This plant has hairs with small balls of a glistening sticky material on the ends. Insects are attracted to this substance and become stuck to the hair. Then the other hairs wrap around the insect, suffocating it and coating it in digestive fluid. The plant absorbs the nutrients or holds the bug captive so other bugs—called "assassin" bugs—can eat it. The plant absorbs the assassin bugs' waste.

Monkey cup Shaped like a pitcher, this plant uses its specialized leaves to collect rainwater. Insects are attracted to the sweet substances on the rim of each leaf. They soon find themselves stuck in the pool of water, unable to climb out because many tiny downward-pointing hairs block the way. The insect eventually drowns and gets digested.

Bladderwort Most of the bladderwort's actions take place underground. The plant usually grows in water or in water-soaked soil. It uses bladders, or sacs, to trap its prey. Once tiny free-swimming organisms bump into the bladders, hairlike organs signal a trap door on the bladder to open. The plant uses water pressure to suck the organisms in, closes the door, and immediately starts digesting them.

Underwater view of a bladderwort

All Choked Up

Another type of plant that survives off other organisms is the strangler fig. Strangler figs exist in the highly competitive rain forests, where it's difficult for plants to find food. To solve this problem, strangler figs start off as air plants, growing on a tree like moss. Eventually, the strangler fig grows roots that surround the tree and reach to the ground to get more nutrients (left). The strangler fig almost completely covers the original tree, which dies because it can't get enough food and water.

Activity

GONE HUNTING Compare and contrast a Venus flytrap and a Monkey cup. What are the similarities and differences between their structures, feeding strategies, and adaptations? Make a Venn diagram to indicate what they have in common, and what features are unique to each.

HEROES
THE Plant Doctor

At the age of seven, George Washington Carver was nicknamed "the plant doctor" by his neighbors. His love of all living things began in a "secret garden" he created in the woods behind his home. "I had never heard of botany and could scarcely read," he recalled years later. People from miles around would bring him their dying plants and he would nurse them back to life.

Carver at work in his greenhouse

Born about 1864, George was an orphan raised by his former slave owners, Moses and Susan Carver, on their farm in Diamond Grove, Missouri. When he was old enough for school, young George moved to the nearby town of Neosho to attend a one-room school for African-American children. There he lived with Mariah and Andrew Watkins. Mrs. Watkins, a nurse and midwife, taught George about plants and herbs that could cure aches and pains. Years later, Carver extracted oil from peanuts; it helped polio patients when massaged into the skin.

While Carver was studying for his master's degree at Iowa State University, the famous educator Booker T. Washington invited him to teach at Tuskegee Normal and Industrial Institute in Alabama. Carver joined the faculty in 1896 as head of the new agricultural school, where he remained for the rest of his career.

"The Wizard of Tuskegee"

At Tuskegee, Carver was put in charge of the Agricultural Experiment Station, a 10-acre plot where he could conduct plant experiments and host farmer's conferences. He published bulletins and, for those who couldn't read, traveled with a demonstration wagon, helping local people run their farms and households more efficiently. He also taught people how to preserve food by canning or drying it.

While visiting cotton farms, Carver noticed that the cotton crops looked sickly. He chemically analyzed soil samples and concluded that the constant planting of cotton robbed the soil of nitrogen, a nutrient needed for plants to grow. He began looking for ways to enrich the soil.

He and his students made an organic fertilizer from rags, string, and other cast-off items. To that he added manure, soil, grass, and leaves. He plowed the land and spread the natural fertilizer all around. Then he planted cowpeas, or legumes. Their pods take nitrogen from the air and use their roots to deposit it in the soil. The result: a better cotton crop and richer soil.

Carver concluded that planting cotton year after year did not allow the soil to rest and replenish its nutrients. The boll weevil, a beetle, was wiping out entire crops of cotton. But he also noticed that the peanut and the sweet potato crops were not affected by the beetle at all.

He encouraged the local farmers to rotate their crops, planting cotton one year, sweet potatoes the next, and peanuts after that. Carver believed that farmers could grow almost anything with proper cultivation of the soil—and he proved it.

Carver examines a tree.

and laundry soap. In 1925 he published a bulletin, *How to Grow the Peanut and 105 Ways of Preparing it for Human Consumption*. It featured recipes for peanut butter, peanut cookies, wafers, cakes, caramels, muffins, ice cream, and many other foods. At Tuskegee, he hosted a luncheon with foods created entirely from the peanut. Carver also went to Washington, D.C., and testified before the House Ways and Means Committee to support the American peanut industry.

Many people offered Carver jobs during his lifetime, including inventors Thomas Edison and Henry Ford. He turned them all down, preferring to stay at Tuskegee, where he set up a foundation and a museum. When George Washington Carver died in 1943, the world of science and all of America mourned the loss of this gifted teacher.

The Peanut Man

In 1919 Carver made a discovery that changed the course of his career. While breaking down the chemical properties of the peanut, he discovered a way to make a kind of milk from peanuts.

By chemically breaking down its sugars, starches, gums, and oils, Carver found more than 300 uses for the peanut, including treatments for polio and diabetes; dyes, paints, skin creams, polishes; diesel fuel;

Activity

CHECK OUT THE SOIL

Go to your local woods and examine the soil. Using a fine shovel, dig several inches of soil and analyze the various layers. Describe what you see and write these findings in a journal. Now do the same in a different location. What changes in the soil do you notice? How is the soil different near different plants?

PICTURE THIS

POLLINATION
TO BEE OR NOT TO BE

A flower's appearance and structure greatly affect how it's pollinated. Pollination begins when pollen grains from the male part of the flower land on the female part. In most cases, the wind or animal pollinators help the process.

WHAT'S THE BUZZ?
Insects like, and even depend on, flowers. Butterflies, moths, and particularly bees get most of their nutrition from the flowers they visit. To attract these guests, some plants produce nectar, a sweet-smelling, sweet-tasting substance. As the insect drinks the nectar, it might pick up bits of pollen dust on its fuzzy body. When it stops at another plant, it leaves some of the pollen behind, and fertilization begins.

A bee pollinates a flower.

SOOOOO SWEET!
To protect their valuable content, flowers that produce nectar have evolved into many shapes. Often the petals of nectar-producing flowers form funnels, bells, and tubes—with the goodies hidden deep inside. Bees, butterflies, and moths eat the nectar; some use their long tube-shaped mouths and long tongues.

TOUGH IT OUT
Mammals get into the pollinating act, too. Some bats pollinate the giant saguaro and other cactus flowers. In tropical areas many plants depend on the long nose and tongue of the bat to pollinate the flowers that produce bananas, avocados, mangoes, and guavas. Bats have poor eyesight, yet they do have strong claws to cling to the plants. This is why the plant species pollinated by bats must be tough and large.

A bat pollinates a cactus flower.

MASQUERADE BALL Some orchids actually look and smell like the female mates of certain wasps and flies. They can even carry the fragrance female insects give off when they are ready to mate. Eager males follow the scent and try their hardest to complete the mating dance. On their next visit to a flower, they shake off the orchid pollen stuck to their bodies!

AH-CHOO! In the summer, when your eyes and nose get to itching, it could be from the pollen grains floating through the air. The flowers of many wind-pollinated plants, such as grasses, are often pale, small, and have no fragrance. These plants depend on the breezes, not animals, to carry on their family tradition.

A microscopic view of pollen particles

SEEING RED Brightly colored flowers appeal to animals that are awake and hungry during the daytime. Bees buzz most around flowers that are blue and yellow, while hummingbirds take a direct route to red waxy flowers (above). Scientists believe birds can see the vivid color red, and insects see a paler version that's not as rosy. Pale petals stand out against the darkness, so flowers pollinated by moths and bats are often very light colored or white.

In a single day, a hummingbird can consume more than twice its own body weight in nectar.

TREES IN THE BREEZE Have you ever seen a helicopter land right at your feet? Chances are you have, especially if you've been around pine trees. Most pine trees carry both male and female cones. Pollen dust is released by the male cone and is carried by the wind. Some of it gets stuck on the female cones, causing the cone to grow to four times its size. Now packed with seeds, the mature cone opens, releasing seeds that have a winglike scale attached. As the seeds fall, they spin through the air like helicopters, eventually landing in an area where a new pine can grow.

A female pine cone

THE PLANT FILES

Older than the Dinosaurs

Once upon a time all the land on Earth was completely bare. There were no plants, trees, or animal life of any kind—only rocks, dirt, and water. Believe it or not, this was our planet about 415 million years ago. And then the first land plants hit the scene.

Moss: The Velvet Carpet

What gives the forest that lush, cool inviting feeling? Think of a blanket of moss beneath your feet. If you see a small green plant only a few centimeters tall, with no flowers, clustered together with hundreds of others just like it, chances are it's a moss (background picture).

Mosses don't reproduce using seeds. They also don't have true roots or leaves, or tissues that conduct water. Instead, mosses soak up water and minerals using hairlike structures called rhizoids, which also anchor them to the soil. Because they don't have roots, mosses usually prefer moist, damp places. Mosses reproduce by dispersing free spores, produced in a capsule at the tip of a long stalk.

Fascinatin' Ferns

Once, 300 million years ago, many ferns grew as tall as 50 feet. Today fern species aren't nearly so big. Ferns do not have seeds and often grow in moist shady areas, like swamps and along stream banks. Some grow on dry ground or rocks, while others grow on the tops of trees in rain forests. Many ferns have underground stems that grow just beneath the soil, but the tree fern has a sturdy treelike trunk.

Many kinds of ferns have fronds, or elaborate leaves. As fronds grow up from the ground, they are tightly coiled. Because they resemble the head of a violin, they are called fiddleheads. Once they uncoil, fiddleheads become mature fronds.

Fiddlehead fern

Tree fern

Check out these other ferns:

Common polypody ferns Polypody means "many feet" in Greek. No wonder: These ferns grow in shady areas and on the tops of rocks on ledges and spread rapidly. Whole ledges of plants all bunched together become a tangled mat of many feet.

Tree ferns What makes these ferns special are their stems, which are as rigid as tree trunks. The largest fern in the world is the Ponga tree in New Zealand, which can grow up to 30 feet (10 m) high.

Whisk fern

When's a Fern Not a Fern?

When it's a fern ally. Whisk ferns, clubmosses, spikemosses, quillworts, and horsetails are all fern allies. These are seedless plants that have water-conducting tissues. Fern allies may seem small and insignificant, but don't be fooled. They are descendants of ancient plants that dominated Earth for many millions of years.

The whisk fern has no true leaves or roots. It reproduces by liberating spores from spore-bearing capsules. Clubmosses are often creeping plants with spores clustered in small cones. The leaves are needle-like or scale-like with a single vein. Fossils indicate that clubmosses once grew 100 feet (30 m) tall!

The spikemoss resembles the clubmoss, but it has cones at the branch tips. Some, such as the resurrection plant, can survive in the desert. When full grown, its branches curl up into a brown ball, dry out, and appear almost dead. Once it gets water, the resurrection plant's leaves unfold and seem—miraculously—to come back to life.

Resurrection plants in their normal (left) and dried-up states

Filmy fern

Filmy Ferns These ferns are so delicate that their leaves are often only a few cells thick.

Activity

LET IT FLOW Ferns have tissues to conduct water, and mosses don't. Try this demonstration with a stalk of celery to see water conduction in action. Add several drops of red food coloring to a glass of water. Put a celery stalk (preferably one with leaves at the top) in the glass. Check the celery stalk several hours later. Observe any changes. What does this experiment tell you about plants and how they conduct water?

MAP
Around the World in 80 Bites

We depend on plants to survive, and so we've learned to grow the foods we need. Scientists estimate that humans began farming about 10,000 BC. These early farmers raised plants that were native to their region. Gradually explorers traveled the world, bringing back new food-producing plants to their homelands. They also brought their native crops to new areas. Crops such as wheat, corn, and soybeans may be found all over the world today, but each one had to start somewhere. Check out the map to learn where some familiar foods were first cultivated.

❶ The Americas: Corn Native Americans first grew corn about 3,000 years ago. They shared their seeds and farming methods with European explorers. Today the United States produces most of the world's supply of popcorn.

❷ Central and South America: Cocoa Beans The cacao tree produces cocoa beans that are processed to make cocoa and chocolate. Cocoa beans were so important to the Aztec rulers of Mexico that they used the beans for food and drink, and as a form of money.

❸ South America: Potatoes Native to the Andes mountains, potatoes grew there as early as 1,800 years ago. They were brought to Ireland much later and became a staple crop in the United States after a large wave of Irish immigrants arrived in the 1800s.

❹ Central America: Tomatoes Spanish and Portuguese explorers first brought tomatoes home in the 15th and 16th centuries. They realized that tomatoes could grow in the warm environment in Spain just as easily as in their original environment.

❺ South America: Peanuts Spanish and Portuguese explorers first brought peanuts to Europe in the 1500s.

6 Western Europe: Oats Begun as a weed in western Europe, the hardy oat plant can grow almost anywhere. Used mainly to feed cattle, oats are also an ingredient in many breakfast cereals.

8 Africa: Wheat One of the first grains to be cultivated over 9,000 years ago, wheat can withstand cold temperatures and has adapted to growing in very cold climates. China is the largest producer of wheat in the world today, while the United States is the third.

9 Southwestern Asia: Onions One of the world's oldest cultivated plants, the onion has long been added to different foods. Traditional healers have applied onion to animal bites, warts, burns, and other injuries.

10 East Asia: Soybeans Inexpensive to harvest, soybeans are a rich source of protein for both humans and animals. Today it is one of the largest crops in the United States.

12 Southern Pacific Islands: Sugar The sugar-producing cane plant grows in very warm climates. Used in a variety of foods around the world, sugarcane is one of the few crops that is still harvested by hand.

11 Asia: Rice Nearly one-half of the world's population eats rice as a staple food. Scientists estimate that rice cultivation began in India around 10,000 BC. Farmers submerge rice seedlings in water in a rice field, or paddy. Rice grows practically anywhere that land can be flooded.

7 Africa: Sorghum The seeds of the sorghum plant are used in many different dishes around the world. The stalk of the sorghum plant is sometimes used to make brooms and brushes. Sorghum can survive drought, which makes this plant important in the world's hottest and driest regions.

Activity

LOCAL CROPS What kinds of crops are grown in your community? Head to your local library or surf the Internet to find this information. Create a map showing local crops and explain why and where these crops are grown. Be sure to include any crops that are no longer grown in your region and discuss why you think this is so.

THE PLANT FILES

SCIENTIST'S NOTEBOOK

Saving Seeds

When Suzanne Nelson was a child, she loved collecting things—not Beanie Babies™ or Pokémon™ cards, but such found treasures as feathers and bits of glimmering glass. Today she collects seeds. To many scientists, grains and other types of seeds are as valuable as gems. Some Russian scientists even starved to death rather than sacrifice samples of grain during World War II.

Nelson is the seed bank curator, or caretaker, at the Native Seeds/SEARCH project in Tucson, Arizona. Instead of money, her bank saves seeds from crops traditionally grown in the southwestern United States and northwestern Mexico. Over many generations, these crops—maize (corn), melon, squash, and other plants—have adapted to the desert and are a precious resource. For thousands of years, these adapted plants have supplied Native Americans with food, medicines, dyes, and basket-making materials. But some seeds, like squash, had become very rare. Native Americans wanted to grow these traditional crops again, but could not locate any seeds, so they turned to Gary Nabhan and Mahina Drees, and Native Seeds/SEARCH (NS/S) began. Today Nelson's seed bank contains almost 2,000 samples from traditional crops and their wild crop relatives. It's vital to preserve crop genetic diversity so that plants will still produce even if confronted with disease, insects, or drought.

"A seed bank needs to be managed very carefully," Nelson says. She must decide which seeds need to be collected. The seed-bearing fruits must be picked at just the right times in just the right places—sometimes that means taking an exhausting hike to a remote site. The fruits and the seeds are cleaned and then photographed, so that scientists will remember what they look like when freshly picked. To make sure the plants sprout, the seeds are dried, frozen inside plastic bags, and tested. Properly stored, the seeds should grow for at least 50 years.

Saving seeds is one project goal. Another is "growing out" the right number of seeds so that new plants will produce enough seeds to be distributed to farmers and home gardeners. NS/S currently

Suzanne Nelson holds a jar of seeds. She works to keep native traditions alive in the Southwest.

Nelson waters the seedlings of 43 chile varieties.

"Wild chiles make a sensational salsa," says Nelson.

distributes several dozen kinds of seeds, including black-eyed peas, cotton, garbanzo beans, indigo (which produces a blue dye), melons, sunflowers, and wild chiles, or chiltepines (CHILL-ta-peens). Believe it or not, Nelson says that wild chiles make a good topping for vanilla ice cream. "It's an extraordinary combination—like cold fire," she reports.

One of the world's northernmost populations of chiltepines flourishes in Arizona's Rock Corral Canyon. Scientists from NS/S are very interested in these indigenous plants because they contain the genes for disease and pest resistance and drought and frost tolerance. These plants are more than 8,000 years old and are only one of 23 types of wild chiles growing around the world. Wild chiltepines are the genetic base for the domestic kinds of chiles. Chiltepines have many uses, including food, insect sprays, and medicines.

For about 10 years, researchers have been studying how grazing affects the Rock Corral Canyon chiltepines, as well as which wild animals eat them, and how their seeds disperse into new areas. In 1999 cattle grazing and fires threatened the plants further, so the canyon was officially designated as the Wild Chile Botanical Area—the first place ever set aside to protect the wild relatives of an important crop.

Several years ago the project came to the rescue of sunflowers, too. At the time, a fungus was spreading and the sunflower crops had no natural resistance. But some of the seeds in the seed bank produced disease-resistant plants. These seeds helped develop plants that sunflower farmers could grow successfully. Thanks to NS/S, the sunflowers survived.

In past generations, saving seeds from the yearly harvest was part of everyday life.

Beans, beans, beans: Ojo de Cabra, Anasazi beans, Hopi purple string beans, scarlet runners, and more

Farmers would set sufficient quantities aside to feed their families in winter and to replant their fields in spring. Running out of seeds could be a major problem. Today, thanks to Suzanne Nelson and others at Native Seeds/SEARCH, farmers can ensure that traditional crops survive.

Activity

FOOD SHORTAGE! Imagine running out of wheat. Go to your local library or grocery store and make a list of the foods that contain wheat. Make a flow chart showing these foods (those made with flour; those that use wheat in its grain form). Then list the foods you would have to eliminate from your diet. Describe how this would affect your food supply and your community.

VIRTUAL VOYAGE
LIVIN' LARGE

IMAGINE that you're an organ pipe—no, not that kind, but a member of the cactus family. Take a trip to Organ Pipe National Monument in the Sonoran Desert of southern Arizona, where these giant cacti grow. There you can really put down some roots.

Like most plants, you'll start out as a seed. You shouldn't be in any hurry, because this trip could take years. And once you arrive, you'll stay in place for a century or more. Be prepared to spend the first year underground. Don't worry—you won't be in too deep.

Before you put your head in the sand, you should start at the beginning—your nursery, where all seeds are born. You'd probably call it a flower. The organ pipe's flowers are white because they attract their visitors mainly at night. Between their bright color and sweet smell, the moths and bats can't leave them alone. The organ pipe cactus needs these night fliers: Without them carrying pollen from flower to flower, the cactus couldn't make new seeds.

As a seed, you may not like being eaten by a bird or winding up in some desert rat's cheek pouch, but this is how a lot of seeds get spread. Once on the ground, you'll stay near the surface to get any moisture that comes

White organ pipe flowers help attract nighttime visitors.

along. There isn't much—they don't call this place a desert for nothing. Out here, you'll get rain only a few times a year. Here comes a spring shower, so drink up while you can!

That water really hit the spot. You may feel tingly now; your roots are spreading. Sure enough, they're growing straight out. Other seeds, the ones that send their roots deep into the ground, may call you shallow. But you can't get too deep. The closer you stay to the surface, the easier it is for your roots to slurp up any water that comes by.

Are you getting bored yet? It's only been a year. But that's how long it takes for you to send up a sprout. Feel the changes: You're losing your hard outer seed coat and putting on more cells. Notice how the cells feel like they're coated with wax? This is to hold water in and keep the Sun from sucking out your moisture.

Now that you're in the sunlight, your cells are dividing as you get bigger. As you grow, you shoot upwards on your brand new stem. Your stem is getting very heavy, and you need some support. Some of your cells make a tough, woody frame, and they'll make the ribs under the waxy skin. You need them to hold you up—you're going to be at least 20 feet tall.

Now that your roots are feeding you, you may also feel a growth spurt coming on. Do you feel like a teenager or what? Some of your cells are acting up. Actually they're acting sideways. Your arms are branching out and up, kind of like the pipes of a church organ. Now you know where the name came from.

Watch out for those spines coming out of you! But then again, you *are* a cactus. Your spines may be a pain, but you absolutely cannot live without them. They keep nasty desert varmints from chewing on your skin to get at your precious water supply inside.

Well, now that you're full-grown, your journey is ended. You've gone from a tiny seed to a cactus the size of a pipe organ. You're welcome to stay as long as you like—but only if you have the spines for it!

A bat can't resist a good bite of cactus.

Organ pipe cactus

Activity

SOAK IT UP Cacti are adapted amazingly well to their desert environments. To see for yourself, get a small cactus plant from a plant store. Remove a small portion of the cactus leaf with scissors—don't touch it directly and wear heavy work gloves. Leave the leaf in the Sun to dry it out. Check it after several hours, and observe it up close. How long does it take to lose its moisture completely?

The Suspicious SPECIMENS

SOLVE-IT-YOURSELF MYSTERY

"Here we are!" said Herb Greene. The camp counselor and his scouts had been hiking all morning. "We'll stop here for the night." Sighs of relief echoed through the campsite as the hikers dropped their heavy backpacks.

"This is a great place to camp," said Herb. "We can explore several different wild environments near here. We're close to a stream, field meadow, forest, and a mountain."

"What's over there?" asked Leif.

"About two miles southeast you'll find a nice open field meadow," said Herb.

"Now I want each of you to head off in a different direction and see what kinds of plants are unique to that environment. Take your field guide and identify as many as you can. Make sure to record them in your notebook, and bring back pictures of leaves and flowers."

"Why don't you get back by three o'clock," continued Herb. "Each of you needs to take a compass, map, and walkie-talkie from the supply bag. I'll be here at the base camp if you need me. Good luck!"

Leif and Hazel followed the path along the stream. Rudy went to the open meadow to find some plants that grow well in the Sun. Laurel and Bud hiked through the woods toward the mountain.

A bit later, Bud returned to base camp to find Herb lighting a campfire.

"Welcome back!" said Herb. "So you're the first to return. What did you find?"

"Well, I saw a deer about 50 feet from me, but it took off before I could snap a photo. I found lots of conifers, like pine, spruce, and fir. Here are their cones," he said, pulling them out of his pack. "And look at this. I've identified it as a Lady's Slipper."

"Yes, those are common in coniferous woods. Good job," said Herb.

A few moments later, Rudy, Leif, and Hazel bounded back into camp at the same time.

"Well," started Leif, "Hazel and I saw lots of ferns along the stream."

"And pretty little water buttercups in the water," Hazel added.

"Well, what did you find, Rudy?" Herb asked.

"I didn't find too much in the field," said Rudy.

"I'm surprised," said Herb.

"But I did find this little salamander crawling around!" Rudy held it up for everyone to see.

"What else besides that little beast?" laughed Herb.

"The meadow was pretty dried out. Mainly grasses and stuff. I did find this clover, though." Rudy handed Herb a small plant he had plucked out of the ground.

Herb looked at the plant and noticed the roots still had some moist dark-brown soil clinging to them.

Forty minutes later the rest of the group finally returned, eager to compare notes with the others.

"I found huckleberry," blurted Laurel.

"You've found some interesting plants," said Herb. "But are all of you certain that you went where you were supposed to go? Let me get my field guide."

Herb looked down at the field guide lying next to the supply bag, which still had a compass in it. "Aha! It looks like someone didn't take a compass. That's proof that one of you didn't go where you should have gone!"

Who was it?

Note: All of the plants described here are native to Oregon.

Clues

Look at the picture below to figure out who wandered in the wrong direction and use these clues . . .

1. Stream runs along the north side of camp and ends about a half mile from camp.
2. Woods are north of camp.
3. Woods are on the north side of the stream.
4. Mountain is about 2 miles north of woods.
5. Field meadow is about 2 miles southeast of camp.

Answer on page 32

WILD PLANTS

FUN & FANTASTIC

Get a Whiff of This!

- In the rain forests of Indonesia, the *Amorphophallus titanum* (right) produces a huge spike of flowers that open and smell like a delicious mixture of rotten fish and burnt sugar. Flies love the scent and come from miles around to lay their eggs on the plant.

- The **baobab tree** in Africa has a trunk so huge that it takes 15 adults with outstretched arms to circle it. Its flowers smell musty and rancid, almost like urine, a scent that appeals to bats. The trunk is made into rope and cloth, and its edible fruit is known as "monkey-bread."

THERE'S A FUNGUS AMONG US!

Because **fungi** don't produce their own food, they are *not* classified as plants. Recent research shows that mushrooms and other fungi are more closely related to animals than plants. Just think—you might be related to a topping on your next pizza.

MAN, ARE YOU OLD!

Ginkgo-saurus

Scientists have found leaf imprints of **ginkgo** tree ancestors in sedimentary rocks dating back to the Jurassic and Triassic periods (about 135–210 million years ago). This means **ginkgo** trees date back to the early dinosaur days.

Joke for a Green Thumb?

Q: How can you identify a dogwood?

A: By its bark.

It's No Ice Cream Cone

In the White-Inyo mountain range of California stands a bristlecone pine named *"Methuselah."* The tree is about 4,765 years old, which means it started as a seed in the United States when the Egyptians began building their pyramids. This makes *Methuselah* the oldest living plant in the world. But don't expect to find it—rangers won't reveal its location.

Foxglove's Legend

The *foxglove* has several names: fairy-caps, fairy-petticoats, and fairy-thimbles. According to legend, fairies would give the *foxglove* blossoms to foxes to wear as gloves. That way they wouldn't get caught trying to steal from the chicken coop.

"The Courier"

Some plants and animals just can't live without each other. The Indian rhino in Nepal lives off the *trewia tree's* fruits and seeds. Some of the plant's seeds pass through the animal's body in its waste. The seeds land on a muddy area near the river bank, the ideal place for *trewia trees* to grow. Unfortunately, this rhinoceros is becoming rare, so the trewia tree may also become rare.

Feed Me!

Seymour, a young assistant in a flower shop, finds that a plant he's growing suddenly turns the shop into a major attraction. The plant starts growing at a rapid rate. It also develops a voice, craves human blood, and cries out "Feed me! Feed me!" *The Little Shop of Horrors*, originally a play and then a musical, was later made into a movie featuring Steve Martin and Rick Moranis.

YOUR WORLD YOUR TURN

Classroom Garden

Recently a study conducted by the Nature Conservancy found that the United States is more biologically diverse than people once thought. The 50 states contain about 10 percent of the known plant species on Earth. We also have a large variety of ecological regions. Of the world's six largest countries, we have the most varied ecological regions, including deserts, forest, grasslands, tundra, and more.

Every year in North America, about 30 previously unknown species of flowering plants are discovered. The same study suggests that many species are extinct or missing, and many are threatened. Consider what the earth would look like with fewer and fewer plants and more and more strip malls!

From a regional field guide, choose 15 plants that grow near your school. Form teams; each team should study one plant. Consider the following:

- What is its scientific name and what family is it from?
- What is its most interesting feature?
- What does it look like? When does it bloom?
- How does it grow?
- What types of animals feed on it, if any?
- What is its habitat?
- How does it reproduce?
- How does it protect itself?
- Does humankind use this plant?
- Is it an edible plant?
- Is this the only type of environment it lives in? If not, where else could you find it?
- How does it adapt to live in this environment?
- Is this plant native to this area or was it introduced within the last 100 years?

Once you have researched your plant, create a large chart showing all 15 plants, with lists of their characteristics. Drawings could accompany each plant. What have you learned about each other's plants?

Try to locate real samples of the plants in the schoolyard or nearby wild or grassy areas. Are any of these plants threatened? If so, what could you do to save them? Write your local nature conservancy to find out what is being done, and what you can do to help.

Ready for the ultimate challenge? Enter this or any other science project in the Discovery Young Scientist Challenge. Visit *discoveryschool.com/dysc* to find out how.

ANSWER
Solve-It-Yourself Mystery, pages 28–29

Rudy said he went to the meadow, but he never stepped foot in it. When Rudy realized how far away the field was, he decided not to go and stayed close to camp to rest. Also, Rudy arrived about the same time as the others who weren't more than a half-mile away. This led Herb to believe that Rudy couldn't have come back so quickly from the meadow, which was two miles away. There he would have been sure to find certain varieties of sunflowers, field buttercups, and clover. Rudy *did* find clover, but it was a variety of clover that he easily plucked out of the damp soil near the stream close to camp. Soil in open fields is exposed to more direct sunlight and heat and tends to be significantly drier. Rudy wouldn't have been able to pluck an entire plant out of the ground if he had gone to the meadow. The delicate roots from a dry field would have looked damaged. Also, it's unlikely that Rudy would have seen a salamander in an open field. He probably caught it while standing by the stream.

32 DISCOVERY CHANNEL SCHOOL